Preface

These supplemental chapters accompany the third edition of *Introduction to the Practice of Statistics (IPS)*. These chapters can also be found electronically on the IPS CD-ROM.

In writing *IPS* we have concentrated on the material we think most valuable in a first course. Our criteria included direct usefulness in practice, and the extent to which studying this material prepares students to go on to more advanced statistical methods. We think that encyclopedic introductory texts intimidate students and that very brief coverage of many methods reinforces students' perceptions that statistics consists of recipes. This print supplement allows us to offer additional (optional) material without violating these principles.

This supplement contains two new chapters complete with exercises. Chapter 14, Nonparametric Tests, presents the most common rank tests: Wilcoxon-Mann-Whitney, Wilcoxon signed rank, and Kruskal-Wallace, along with comments on their use in practice. Chapter 15, Logistic Regression, introduces a prominent topic in statistical practice that we would like to see presented more often in beginning instruction.

CHAPTER **14**

Nonparametric Tests

Introduction
14.1 The Wilcoxon Rank Sum Test
14.2 The Wilcoxon Signed Rank Test
14.3 The Kruskal-Wallis Test

Introduction

The most commonly used methods for inference about the means of quantitative response variables assume that the variables in question have normal distributions in the population or populations from which we draw our data. In practice, of course, no distribution is exactly normal. Fortunately, our usual methods for inference about population means (the one-sample and two-sample t procedures and analysis of variance) are quite **robust.** That is, the results of inference are not very sensitive to moderate lack of normality, especially when the samples are reasonably large. We gave some practical guidelines for taking advantage of the robustness of these methods in Chapter 7.

What can we do if normal quantile plots suggest that the data are clearly not normal, especially when we have only a few observations? This is not a simple question. Here are the basic options:

robustness

outliers

1. If there are extreme **outliers** in a small data set, any inference method may be suspect. An outlier is an observation that may not come from the same population as the others. To decide what to do, you must find the cause of the outlier. Equipment failure that produced a bad measurement, for example, entitles you to remove the outlier and analyze the remaining data. If the outlier appears to be "real data," it is risky to draw any conclusion from just a few observations. This is the advice we gave to the child development researcher in Example 2.19 (page 163).

transforming data

2. Sometimes we can **transform** our data so that their distribution is more nearly normal. Transformations such as the logarithm that pull in the long tail of right-skewed distributions are particularly helpful. We used the logarithm transformation in Example 7.10 (page 519) to make the right-skewed distribution of carbon monoxide in vehicle exhausts more nearly normal.

3. In some settings, **other standard distributions** replace the normal distributions as models for the overall pattern in the population. We mentioned in Section 5.2 that the Weibull distributions are common models for the lifetimes in service of equipment in statistical studies of reliability. There are inference procedures for the parameters of these distributions that replace the t procedures when we use specific nonnormal models.

nonparametric methods

4. Finally, there are inference procedures that do not assume any specific form for the distribution of the population. These are called **nonparametric methods.** They are the subject of this chapter.

The word *nonparametric* contrasts these methods with statistical methods that are based on models of a specific form and use data to estimate the parameters in these models. For example, simple linear regression (Sections 2.3 and 10.1) uses a straight-line model. The parameters in this model are the

Setting	Normal test	Rank test
One sample	One-sample t test Section 7.1	Wilcoxon signed rank test Section 14.2
Matched pairs	Apply one-sample test to differences within pairs	
Two independent samples	Two-sample t test Section 7.2	Wilcoxon rank sum test Section 14.1
Several independent samples	One-way ANOVA F test Chapter 12	Kruskal-Wallis test Section 14.3

FIGURE 14.1 Comparison of tests based on normal distributions with nonparametric tests for similar settings.

nonparametric regression

slope and intercept of the line, and we can use the least-squares method to estimate these parameters from data. Scatterplot smoothers (Section 2.1), in contrast, do not assume any specific form for the relationship. When we use a smoother, we are doing **nonparametric regression.**

This chapter concerns one type of nonparametric procedure, tests that can replace the t tests and one-way analysis of variance when the normality assumptions for those tests are not met. There are two big ideas that can serve as the basis for nonparametric tests. One is to use *counts*. This is the basis of the **sign test** for matched pairs, discussed on pages 519–522. The other is to use *ranks*. This chapter discusses rank tests.

sign test

Figure 14.1 presents an outline of the standard tests (based on normal distributions) and the rank tests that compete with them. All of these tests concern the *center* of a population or populations. When a population has at least roughly a normal distribution, we describe its center by the mean. The "normal tests" in Figure 14.1 all test hypotheses about population means. When distributions are strongly skewed, the mean may not be the preferred measure of center. We will see that rank tests do not test hypotheses about means.

We devote a section of this chapter to each of the rank procedures. Section 14.1, which discusses the most common of these tests, also contains general information about rank tests. The kind of assumptions required, the nature of the hypotheses tested, the big idea of using ranks, and the contrast between exact distributions for use with small samples and approximations for use with larger samples are common to all rank tests. Sections 14.2 and 14.3 more briefly describe other rank tests.

14.1 The Wilcoxon Rank Sum Test

Two-sample problems (see page 537) are among the most common in statistics. The most useful nonparametric significance test compares two distributions. Here is an example of this setting.

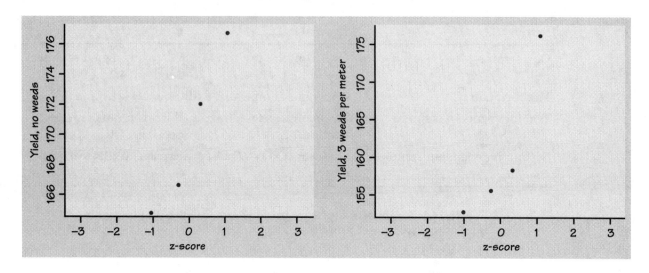

FIGURE 14.2 Normal quantile plots of corn yields from plots with no weeds (left) and with 3 weeds per meter of row (right).

EXAMPLE 14.1 Does the presence of small numbers of weeds reduce the yield of corn? Lamb's-quarter is a common weed in corn fields. A researcher planted corn at the same rate in 8 small plots of ground, then weeded the corn rows by hand to allow no weeds in 4 randomly selected plots and exactly 3 lamb's-quarter plants per meter of row in the other 4 plots. Here are the yields of corn (bushels per acre) in each of the plots:[1]

Weeds per meter	Yield (bushels/acre)			
0	166.7	172.2	165.0	176.9
3	158.6	176.4	153.1	156.0

Normal quantile plots (Figure 14.2) suggest that the data may be right-skewed. The samples are too small to assess normality adequately or to rely on the robustness of the two-sample t test. We may prefer to use a test that does not require normality.

The rank transformation

We first rank all 8 observations together. To do this, arrange them in order from smallest to largest:

| 153.1 | 156.0 | 158.6 | **165.0** | **166.7** | **172.2** | 176.4 | **176.9** |

The boldface entries in the list are the yields with no weeds present. We see that four of the five highest yields come from that group, suggesting that yields

14.1 The Wilcoxon Rank Sum Test

are higher with no weeds. The idea of rank tests is to look just at position in this ordered list. To do this, replace each observation by its order, from 1 (smallest) to 8 (largest). These numbers are the *ranks*:

Yield	153.1	156.0	158.6	**165.0**	**166.7**	172.2	176.4	**176.9**
Rank	1	2	3	**4**	**5**	6	7	**8**

> **Ranks**
>
> To rank observations, first arrange them in order from smallest to largest. The **rank** of each observation is its position in this ordered list, starting with rank 1 for the smallest observation.

Moving from the original observations to their ranks is a transformation of the data, like moving from the observations to their logarithms. The rank transformation retains only the ordering of the observations and makes no other use of their numerical values. Working with ranks allows us to dispense with specific assumptions about the shape of the distribution, such as normality.

The Wilcoxon rank sum test

If the presence of weeds reduces corn yields, we expect the ranks of the yields from plots with weeds to be smaller as a group than the ranks from plots without weeds. We might compare the *sums* of the ranks from the two treatments:

Treatment	Sum of ranks
No weeds	23
Weeds	13

These sums measure how much the ranks of the weed-free plots as a group exceed those of the weedy plots. In fact, the sum of the ranks from 1 to 8 is always equal to 36, so it is enough to report the sum for one of the two groups. If the sum of the ranks for the weed-free group is 23, the ranks for the other group must add to 13 because 23 + 13 = 36. If the weeds have no effect, we would expect the sum of the ranks in either group to be 18 (half of 36). Here are the facts we need in a more general form that takes account of the fact that our two samples need not be the same size.

> **The Wilcoxon rank sum test**
>
> Draw an SRS of size n_1 from one population and draw an independent SRS of size n_2 from a second population. There are N observations in all, where $N = n_1 + n_2$. Rank all N observations. The sum W of the ranks for the first sample is the **Wilcoxon rank sum statistic.** If the two populations have the same continuous distribution, then W has mean
>
> $$\mu_W = \frac{n_1(N+1)}{2}$$
>
> and standard deviation
>
> $$\sigma_W = \sqrt{\frac{n_1 n_2 (N+1)}{12}}$$
>
> The **Wilcoxon rank sum test** rejects the hypothesis that the two populations have identical distributions when the rank sum W is far from its mean.*

In the corn yield study of Example 14.1, we want to test

H_0: no difference in distribution of yields

against the one-sided alternative

H_a: yields are systematically higher in weed-free plots

Our test statistic is the rank sum $W = 23$ for the weed-free plots.

EXAMPLE 14.2 In Example 14.1, $n_1 = 4$, $n_2 = 4$, and there are $N = 8$ observations in all. The sum of ranks for the weed-free plots has mean

$$\mu_W = \frac{n_1(N+1)}{2} = \frac{(4)(9)}{2} = 18$$

and standard deviation

$$\sigma_W = \sqrt{\frac{n_1 n_2 (N+1)}{12}} = \sqrt{\frac{(4)(4)(9)}{12}} = \sqrt{12} = 3.464$$

Although the observed rank sum $W = 23$ is higher than the mean, it is only about 1.4 standard deviations high. We now suspect that the data do not give strong evidence that yields are higher in the population of weed-free corn.

The P-value for our one-sided alternative is $P(W \geq 23)$, the probability that W is at least as large as the value for our data when H_0 is true.

---

*This test was invented by Frank Wilcoxon (1892–1965) in 1945. Wilcoxon was a chemist who encountered statistical problems in his work at the research laboratories of the American Cyanimid company.

To calculate the P-value $P(W \geq 23)$, we need to know the sampling distribution of the rank sum W when the null hypothesis is true. This distribution depends on the two sample sizes n_1 and n_2. Tables are therefore a bit unwieldy, though you can find them in handbooks of statistical tables. Most statistical software will give you P-values, as well as carry out the ranking and calculate W. However, some software packages give only approximate P-values. You must learn what your software offers.

EXAMPLE 14.3 Figure 14.3 shows the output from a software package that calculates the exact sampling distribution of W. We see that the sum of the ranks in the weed-free group is $W = 23$, with P-value $P = 0.10$ against the one-sided alternative that weed-free plots have higher yields. There is some evidence that weeds reduce yield, considering that we have data from only four plots for each treatment. The evidence does not, however, reach the levels usually considered convincing.

It is worth noting that the two-sample t test gives essentially the same result as the Wilcoxon test in Example 14.3 ($t = 1.554$, $P = 0.0937$). It is in fact somewhat unusual to find a strong disagreement between the conclusions reached by these two tests.

The normal approximation

The rank sum statistic W becomes approximately normal as the two sample sizes increase. We can then form yet another z statistic by standardizing W:

$$z = \frac{W - \mu_W}{\sigma_W}$$
$$= \frac{W - n_1(N + 1)/2}{\sqrt{n_1 n_2 (N + 1)/12}}$$

continuity correction Use standard normal probability calculations to find P-values for this statistic. Because W takes only whole-number values, the **continuity correction** improves the accuracy of the approximation.

```
            Exact Wilcoxon rank-sum test

data:   0weeds and 3weeds

rank-sum statistic W = 23, n = 4, m = 4, p-value = 0.100

alternative hypothesis: true mu is greater than 0
```

FIGURE 14.3 Output from the S-Plus statistical software for the data in Example 14.1. This program uses the exact distribution for W when the samples are small and there are no ties (that is, when all observations have different values).

EXAMPLE 14.4 | The standardized rank sum statistic W in our corn yield example is

$$z = \frac{W - \mu_W}{\sigma_W} = \frac{23 - 18}{3.464} = 1.44$$

We expect W to be larger when the alternative hypothesis is true, so the approximate P-value is

$$P(Z \geq 1.44) = 0.0749$$

The continuity correction (see page 386) acts as if the whole number 23 occupies the entire interval from 22.5 to 23.5. We calculate the P-value $P(W \geq 23)$ as $P(W \geq 22.5)$ because the value 23 is included in the range whose probability we want. Here is the calculation:

$$P(W \geq 22.5) = P\left(\frac{W - \mu_W}{\sigma_W} \geq \frac{22.5 - 18}{3.464}\right)$$
$$= P(Z \geq 1.30)$$
$$= 0.0968$$

The continuity correction gives a result closer to the exact value $P = 0.10$.

We recommend always using either the exact distribution (from software or tables) or the continuity correction for the rank sum statistic W. The exact distribution is of course safer for small samples. As Example 14.4 illustrates, however, the normal approximation with the continuity correction is often adequate.

EXAMPLE 14.5
Mann–Whitney test

Figure 14.4 shows the output for our data from two more statistical programs. Minitab offers only the normal approximation, and it refers to the **Mann-Whitney test.** This is an alternate form of the Wilcoxon rank sum test. SAS carries out both the exact and approximate tests. SAS calls the rank sum S rather than W and gives the mean 18 and standard deviation 3.464 as well as the z statistic 1.299 (using the continuity correction). SAS gives the approximate two-sided P-value as 0.1939, so the one-sided result is half this, $P = 0.0970$. This agrees with Minitab and (up to a small roundoff error) with our result in Example 14.4. This approximate P-value is close to the exact result $P = 0.1000$, given by SAS and in Figure 14.3.

What hypothesis does Wilcoxon test?

Our null hypothesis is that weeds do not affect yield. Our alternative hypothesis is that yields are lower when weeds are present. If we are willing to assume that yields are normally distributed, or if we have reasonably large samples, we use the two-sample t test for means. Our hypotheses then become

$$H_0: \mu_1 = \mu_2$$
$$H_a: \mu_1 > \mu_2$$

When the distributions may not be normal, we might restate the hypotheses in terms of population medians rather than means:

$$H_0: \text{median}_1 = \text{median}_2$$
$$H_a: \text{median}_1 > \text{median}_2$$

14.1 The Wilcoxon Rank Sum Test

Mann-Whitney Confidence Interval and Test

```
0 weeds    N =  4   Median =       169.45
3 weeds    N =  4   Median =       157.30
Point estimate for ETA1- ETA2 is      11.30
97.0 Percent C.I. for ETA1- ETA2 is (−11.40,23.80)
W = 23.0
Test of ETA1 = ETA2 vs. ETA1 > ETA2 is significant at 0.0970
```

(a)

```
            Wilcoxon Scores (Rank Sums) for Variable YIELD
                    Classified by Variable WEEDS

                    Sum of        Expected      Std Dev           Mean
WEEDS        N      Scores        Under H0      Under H0          Score

0            4      23.0          18.0          3.46410162        5.75000000
3            4      13.0          18.0          3.46410162        3.25000000


     Wilcoxon 2-Sample Test       S =   23.0000

         Exact P-Values
         (One-sided)   Prob >= S              = 0.1000
         (Two-sided)   Prob >= |S - Mean|     = 0.2000

     Normal Approximation (with Continuity Correction of .5)
         Z =  1.29904      Prob > |Z| = 0.1939
```

(b)

FIGURE 14.4 Output from the Minitab and SAS statistical software for the data in Example 14.1. (a) Minitab uses the normal approximation for the distribution of W. (b) SAS gives both the exact and approximate values.

The Wilcoxon rank sum test provides a significance test for these hypotheses, but only if an additional assumption is met: both populations must have continuous distributions of *the same shape*. That is, the density curve for corn yields with 3 weeds per meter looks exactly like that for no weeds except that it may slide to a different location on the scale of yields. The Minitab output in Figure 14.4(a) states the hypotheses in terms of population medians (which it calls "eta") and also gives a confidence interval for the difference between the two population medians.

The same-shape assumption is too strict to be reasonable in practice. Fortunately, the Wilcoxon test also applies in a much more general and more useful setting. It tests hypotheses that we can state in words as

H_0: two distributions are the same
H_a: one has values that are systematically larger

Here is a more exact statement of the "systematically larger" alternative hypothesis. Take X_1 to be corn yield with no weeds and X_2 to be corn yield with 3 weeds per meter. These yields are random variables. That is, every time we plant a plot with no weeds, the yield is a value of the variable X_1. The probability that the yield is more than 160 bushels per acre when no weeds are present is $P(X_1 > 160)$. If weed-free yields are "systematically larger" than those with weeds, yields higher than 160 should be more likely with no weeds. That is, we should have

$$P(X_1 > 160) > P(X_2 > 160)$$

The alternative hypothesis says that this inequality holds not just for 160 but for *any* yield we care to specify. No weeds always puts more probability "to the right" of whatever yield we are interested in.[2]

This exact statement of the hypotheses we are testing is a bit awkward. The hypotheses really are "nonparametric" because they do not involve any specific parameter such as the mean or median. If the two distributions do have the same shape, the general hypotheses reduce to comparing medians. Many texts and computer outputs state the hypotheses in terms of medians, sometimes ignoring the same-shape requirement. We recommend that you express the hypotheses in words rather than symbols. "Yields are systematically higher in weed-free plots" is easy to understand and is a good statement of the effect that the Wilcoxon test looks for.

Ties

The exact distribution for the Wilcoxon rank sum is obtained assuming that all observations in both samples take different values. This allows us to rank them all. In practice, however, we often find observations tied at the same value. What shall we do? The usual practice is to *assign all tied values the* **average** *of the ranks they occupy.* Here is an example with 6 observations:

average ranks

Observation	153	155	158	158	161	164
Rank	1	2	3.5	3.5	5	6

14.1 The Wilcoxon Rank Sum Test

The tied observations occupy the third and fourth places in the ordered list, so they share rank 3.5.

The exact distribution for the Wilcoxon rank sum W only applies to data without ties. Moreover, the standard deviation σ_W must be adjusted if ties are present. The normal approximation can be used after the standard deviation is adjusted. Statistical software will detect ties, make the necessary adjustment, and switch to the normal approximation. In practice, software is required if you want to use rank tests when the data contain tied values.

It is sometimes useful to apply rank tests to data that have very many ties because the scale of measurement has only a few values. Here is an example.

EXAMPLE 14.6 Food sold at outdoor fairs and festivals may be less safe than food sold in restaurants because it is prepared in temporary locations and often by volunteer help. What do people who attend fairs think about the safety of the food served? One study asked this question of people at a number of fairs in the Midwest:

> How often do you think people become sick because of food they consume prepared at outdoor fairs and festivals?

The possible responses were

    1 = very rarely
    2 = once in a while
    3 = often
    4 = more often than not
    5 = always

In all, 303 people answered the question. Of these, 196 were women and 107 were men. Is there good evidence that men and women differ in their perceptions about food safety at fairs?[3]

We should first ask if the subjects in Example 14.6 are a random sample of people who attend fairs, at least in the Midwest. The researcher visited 11 different fairs. She stood near the entrance and stopped every 25th adult who passed. Because no personal choice was involved in choosing the subjects, we can reasonably treat the data as coming from a random sample. (As usual, there was some nonresponse, which could create bias.)

Here are the data, presented as a two-way table of counts:

	Response					
	1	2	3	4	5	Total
Female	13	108	50	23	2	196
Male	22	57	22	5	1	107
Total	35	165	72	28	3	303

Comparing row percentages shows that the women in the sample are more concerned about food safety than the men:

	Response					
	1	2	3	4	5	Total
Female	6.6%	55.1%	25.5%	11.7%	1.0%	100%
Male	20.6%	53.3%	20.6%	4.7%	1.0%	100%

Is the difference between the genders statistically significant?

We might apply the chi-square test (Chapter 9). It is highly significant ($X^2 = 16.120$, df = 4, $P = 0.0029$). Although the chi-square test answers our general question, it ignores the ordering of the responses and so does not use all of the available information. We would really like to know whether men or women are more concerned about food safety. This question depends on the ordering of responses from least concerned to most concerned. We can use the Wilcoxon test for the hypotheses

H_0: men and women do not differ in their responses
H_a: one of the two genders gives systematically larger responses than the other

The alternative hypothesis is two-sided. Because the responses can take only five values, there are very many ties. All 35 people who chose "very rarely" are tied at 1, and all 165 who chose "once in a while" are tied at 2.

EXAMPLE 14.7   Figure 14.5 gives computer output for the Wilcoxon test. The rank sum for men (using average ranks for ties) is $W = 14{,}112.5$. The standardized value is $z = -3.259$ with two-sided P-value $P = 0.0011$. There is very strong evidence of a difference. Women are more concerned than men about the safety of food served at fairs.

```
           Wilcoxon Scores (Rank Sums) for Variable SFAIR
                   Classified by Variable GENDER

                    Sum of       Expected      Std Dev        Mean
GENDER      N       Scores       Under H0      Under H0       Score

Female      196     31943.5000    29792.0      660.006327    162.977041
Male        107     14112.5000    16264.0      660.006327    131.892523
                Average Scores Were Used for Ties

Wilcoxon 2-Sample Test (Normal Approximation)
(with Continuity Correction of .5)

S =    14112.5   Z = -3.25906   Prob > |Z| = 0.0011
```

FIGURE 14.5 Output from SAS for the food safety study in Example 14.6. The approximate two-sided P-value is 0.0011.

With more than 100 observations in each group and no outliers, we might use the two-sample t even though responses take only five values. In fact, the results for Example 14.6 are $t = 3.2719$ with $P = 0.0012$. The P-value for the two-sample t test is almost exactly equal to that for the Wilcoxon test. There is, however, another reason to prefer the rank test in this example. The t statistic treats the response values 1 through 5 as meaningful numbers. In particular, the possible responses are treated as though they are equally spaced. The difference between "very rarely" and "once in a while" is the same as the difference between "once in a while" and "often." This may not make sense. The rank test, on the other hand, uses only the order of the responses, not their actual values. The responses are arranged in order from least to most concerned about safety, so the rank test makes sense. Some statisticians avoid using t procedures when there is not a fully meaningful scale of measurement.

Limitations of nonparametric tests

The examples we have given illustrate the potential usefulness of nonparametric tests. Nonetheless, rank tests are of secondary importance relative to inference procedures based on the normal distribution.

- Nonparametric inference is largely restricted to simple settings. Normal inference extends to methods for use with complex experimental designs and multiple regression, but nonparametric tests do not. We stress normal inference in part because it leads on to more advanced statistics.

- Normal tests compare means and are accompanied by simple confidence intervals for means or differences between means. When we use nonparametric tests to compare medians, we can also give confidence intervals, though they are rather awkward to calculate. However, the usefulness of nonparametric tests is clearest in settings when they do not simply compare medians—see the discussion of "What hypotheses does Wilcoxon test?" In these settings, there is no measure of the *size* of the observed effect that is closely related to the rank test of the *statistical significance* of the effect.

- The robustness of normal tests for means implies that we rarely encounter data that require nonparametric procedures to obtain reasonably accurate P-values. The t and W tests give very similar results in our examples. Nonetheless, many statisticians would not use a t test in Example 14.6 because the response variable gives only the order of the responses.

- There are more modern and more effective ways to escape the assumption of normality, such as bootstrap methods (see page 445).

SUMMARY

Nonparametric tests do not require any specific form for the distribution of the population from which our samples come.

Rank tests are nonparametric tests based on the **ranks** of observations, their positions in a list ordered from smallest (rank 1) to largest. Tied observations receive the average of their ranks.

The **Wilcoxon rank sum test** compares two distributions to assess whether one has systematically larger values than the other. The Wilcoxon test is based on the **Wilcoxon rank sum statistic** W, which is the sum of the ranks of one of the samples. The Wilcoxon test can replace the **two-sample** t **test.**

P**-values** for the Wilcoxon test are based on the sampling distribution of the rank sum statistic W when the null hypothesis (no difference in distributions) is true. You can find P-values from special tables, software, or a normal approximation (with continuity correction).

SECTION 14.1 EXERCISES

Statistical software is very helpful in doing these exercises. If you do not have access to software, base your work on the normal approximation with continuity correction.

14.1 A study of early childhood education asked kindergarten students to tell a fairy tale that had been read to them earlier in the week. The 10 children in the study included 5 high-progress readers and 5 low-progress readers. Each child told two stories. Story 1 had been read to them; Story 2 had been read and also illustrated with pictures. An expert listened to a recording of the children and assigned a score for certain uses of language. Here are the data (provided by Susan Stadler, Purdue University):

Child	Progress	Story 1 score	Story 2 score
1	high	0.55	0.80
2	high	0.57	0.82
3	high	0.72	0.54
4	high	0.70	0.79
5	high	0.84	0.89
6	low	0.40	0.77
7	low	0.72	0.49
8	low	0.00	0.66
9	low	0.36	0.28
10	low	0.55	0.38

Is there evidence that the scores of high-progress readers are higher than those of low-progress readers when they retell a story they have heard without pictures (Story 1)?

(a) Make normal quantile plots for the 5 responses in each group. Are any major deviations from normality apparent?

(b) Carry out a two-sample t test. State hypotheses and give the two sample means, the t statistic and its P-value, and your conclusion.

(c) Carry out the Wilcoxon rank sum test. State hypotheses and give the rank sum W for high-progress readers, its P-value, and your conclusion. Do the t and Wilcoxon tests lead you to different conclusions?

14.2 Repeat the analysis of Exercise 14.1 for the scores when children retell a story they have heard and seen illustrated with pictures (Story 2).

14.3 Use the data in Exercise 14.1 for children telling Story 2 to carry out by hand the steps in the Wilcoxon rank sum test.

(a) Arrange the 10 observations in order and assign ranks. There are no ties.

(b) Find the rank sum W for the five high-progress readers. What are the mean and standard deviation of W under the null hypothesis that low-progress and high-progress readers do not differ?

(c) Standardize W to obtain a z statistic. Do a normal probability calculation with the continuity correction to obtain a one-sided P-value.

(d) The data for Story 1 contain tied observations. What ranks would you assign to the 10 scores for Story 1?

14.4 The corn yield study of Example 14.1 also examined yields in four plots having 9 lamb's-quarter plants per meter of row. The yields (bushels per acre) in these plots were

162.8	142.4	162.7	162.4

There is a clear outlier, but rechecking the results found that this is the correct yield for this plot. The outlier makes us hesitant to use t procedures because $\bar{x}$ and s are not resistant.

(a) Is there evidence that 9 weeds per meter reduces corn yields when compared with weed-free corn? Use the Wilcoxon rank sum test with the data above and some of the data from Example 14.1 to answer this question.

(b) Compare the results from (a) with those from the two-sample t test for these data.

(c) Now remove the low outlier 142.4 from the data with 9 weeds per meter. Repeat both the Wilcoxon and t analyses. By how much did the outlier reduce the mean yield in its group? By how much did it increase the standard deviation? Did it have a practically important impact on your conclusions?

14.5 Example 7.17 (page 547) reports the results of a study of the effect of the pesticide DDT on nerve activity in rats. The data for the DDT group are

12.207	16.869	25.050	22.429	8.456	20.589

The control group data are

11.074	9.686	12.064	9.351	8.182	6.642

It is difficult to assess normality from such small samples, so we might use a nonparametric test to assess whether DDT affects nerve response.

(a) State the hypotheses for the Wilcoxon test.

(b) Carry out the test. Report the rank sum W, its P-value, and your conclusion.

(c) The two-sample t test used in Example 7.17 found that $t = 2.9912$, $P = 0.0247$. Are your results different enough to change the conclusion of the study?

14.6 In Example 7.14, we compared the DRP scores of two groups of third graders who followed different reading curricula. The data appear in Table 7.2 (page 543).

(a) Apply the Wilcoxon rank sum test to these data and compare your result with the $P = 0.0132$ obtained from the two-sample t test in Example 7.14.

(b) What are the null and alternative hypotheses for the t test? For the Wilcoxon test?

14.7 Table 7.3 (page 552) gives data on blood pressure before and after treatment for two groups of black males. One group took a calcium supplement, and the other group received a placebo. Example 7.20 compares the decrease in blood pressure in the two groups using pooled two-sample t procedures, and Exercise 7.69 applies the more general two-sample t procedures. The normal quantile plot for the calcium group (Figure 7.14, page 552) shows some departure from normality, though not enough to prevent use of t procedures.

(a) Use the Wilcoxon rank sum test to analyze these data. Compare your findings with those of Example 7.20 (page 553) and Exercise 7.69 (page 563).

(b) What are the null and alternative hypotheses for each of the three tests we have applied to these data?

(c) What must we assume about the data to apply each of the three tests?

14.8 Exercise 7.51 (page 556) studies the effect of piano lessons on the spatial-temporal reasoning of preschool children. The data there concern 34 children who took piano lessons and a control group of 44 children. The data take only small whole-number values. Use the Wilcoxon rank sum test (there are many ties) to decide whether piano lessons improve spatial-temporal reasoning.

14.9 Example 14.6 describes a study of the attitudes of people attending outdoor fairs about the safety of the food served at such locations. The full data set is stored on the *IPS* CD as the file eg14.06.dat. It contains the responses of 303 people to several questions. The variables in this data set are (in order)

14.2 The Wilcoxon Signed Rank Test

subject	hfair	sfair	sfast	srest	gender

The variable "sfair" contains the responses described in the example concerning safety of food served at outdoor fairs. The variable "srest" contains responses to the same question asked about food served in restaurants. The variable "gender" contains 1 if the respondent is a woman, 2 if he is a man. We saw that women are more concerned than men about the safety of food served at fairs. Is this also true for restaurants?

14.10 The data file used in Example 14.6 and Exercise 14.9 contains 303 rows, one for each of the 303 respondents. Each row contains the responses of one person to several questions. We wonder if people are more concerned about the safety of food served at fairs than they are about the safety of food served at restaurants. Explain carefully why we *cannot* answer this question by applying the Wilcoxon rank sum test to the variables "sfair" and "srest."

14.11 Shopping at secondhand stores is becoming more popular and has even attracted the attention of business schools. To study customers' attitudes toward secondhand stores, researchers interviewed samples of shoppers at two secondhand stores of the same chain in two cities. Here are data on the incomes of shoppers at the two stores, presented as a two-way table of counts. (From William D. Darley, "Store-choice behavior for pre-owned merchandise," *Journal of Business Research*, 27 (1993), pp. 17–31.)

Income code	Income	City 1	City 2
1	Under $10,000	70	62
2	$10,000 to $19,999	52	63
3	$20,000 to $24,999	69	50
4	$25,000 to $34,999	22	19
5	$35,000 or more	28	24

(a) Is there a relationship between city and income? Use the chi-square test to answer this question.

(b) The chi-square test ignores the ordering of the income categories. The data file ex14.11.dat on the *IPS* CD contains data on the 459 shoppers in this study. The first variable is the city (City1 or City2) and the second is the income code as it appears in the table above (1 to 5). Is there good evidence that shoppers in one city have systematically higher incomes than in the other?

14.2 The Wilcoxon Signed Rank Test

We use the one-sample *t* procedures for inference about the mean of one population or for inference about the mean difference in a matched pairs setting. The matched pairs setting is more important because good studies are generally comparative. The **sign test** (page 519) is a nonparametric test based on counts for matched pairs. We will now meet a rank test for this setting.

EXAMPLE 14.8

A study of early childhood education asked kindergarten students to tell a fairy tale that had been read to them earlier in the week. Each child told two stories. The first had been read to them and the second had been read but also illustrated with pictures. An expert listened to a recording of the children and assigned a score for certain uses of language. Here are the data for five "low-progress" readers in a pilot study:[4]

Child	1	2	3	4	5
Story 2	0.77	0.49	0.66	0.28	0.38
Story 1	0.40	0.72	0.00	0.36	0.55
Difference	0.37	−0.23	0.66	−0.08	−0.17

We wonder if illustrations improve how the children retell a story. We would like to test the hypotheses

H_0: scores have the same distribution for both stories
H_a: scores are systematically higher for Story 2

Because this is a matched pairs design, we base our inference on the differences. The matched pairs t test gives $t = 0.635$ with one-sided P-value $P = 0.280$. Displays of the data (Figure 14.6) suggest a mild lack of normality. We would like to use a rank test.

Positive differences in Example 14.8 indicate that the child performed better telling Story 2. If scores are generally higher with illustrations, the positive differences should be farther from zero in the positive direction than the negative differences are in the negative direction. We therefore compare the

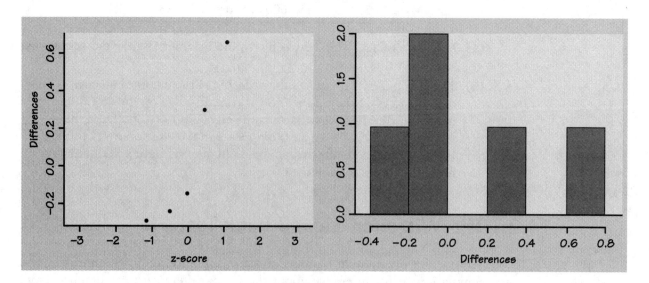

FIGURE 14.6 Normal quantile plot and histogram for the five differences in Example 14.8.

14.2 The Wilcoxon Signed Rank Test

absolute value

absolute values of the differences, that is, their magnitudes without a sign. Here they are, with boldface indicating the positive values:

| **0.37** | 0.23 | **0.66** | 0.08 | 0.17 |

Arrange these in increasing order and assign ranks, keeping track of which values were originally positive. Tied values receive the average of their ranks. If there are zero differences, discard them before ranking.

Absolute value	0.08	0.17	0.23	**0.37**	**0.66**
Rank	1	2	3	**4**	**5**

The test statistic is the sum of the ranks of the positive differences. (We could equally well use the sum of the ranks of the negative differences.) This is the *Wilcoxon signed rank statistic.* Its value here is $W^+ = 9$.

The Wilcoxon signed rank test for matched pairs

Draw an SRS of size n from a population for a matched pairs study and take the differences in responses within pairs. Rank the absolute values of these differences. The sum W^+ of the ranks for the positive differences is the **Wilcoxon signed rank statistic.** If the responses have a continuous distribution that is not affected by the different treatments within pairs, then W^+ has mean

$$\mu_{W^+} = \frac{n(n+1)}{4}$$

and standard deviation

$$\sigma_{W^+} = \sqrt{\frac{n(n+1)(2n+1)}{24}}$$

The **Wilcoxon signed rank test** rejects the hypothesis that there are no systematic differences within pairs when the rank sum W^+ is far from its mean.

EXAMPLE 14.9

In the storytelling study of Example 14.8, $n = 5$. If the null hypothesis (no systematic effect of illustrations) is true, the mean of the signed rank statistic is

$$\mu_{W^+} = \frac{n(n+1)}{4} = \frac{(5)(6)}{4} = 7.5$$

Our observed value $W^+ = 9$ is only slightly larger than this mean. The one-sided P-value is $P(W^+ \geq 9)$.

Figure 14.7 displays the output of two statistical programs. We see that the one-sided P-value for the Wilcoxon signed rank test with $n = 5$ observations and $W^+ = 9$ is $P = 0.4062$. This result differs from the t test result $P = 0.280$, but both tell us that this very small sample gives no evidence that seeing illustrations improves the storytelling of low-progress readers.

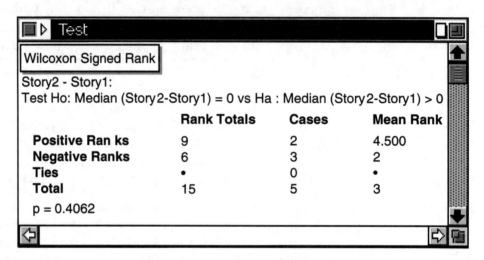

```
            Exact Wilcoxon signed-rank test

data:  Story2-Story1

signed-rank statistic V = 9, n = 5, p-value = 0.4062

alternative hypothesis: true mu is greater than 0
```

FIGURE 14.7 Output from (a) S-Plus and (b) Data Desk for the storytelling study in Example 14.9. These programs use the exact distribution of W^+ when the sample size is small and there are no ties.

The normal approximation

The distribution of the signed rank statistic when the null hypothesis (no difference) is true becomes approximately normal as the sample size becomes large. We can then use normal probability calculations (with the continuity correction) to obtain approximate P-values for W^+. Let's see how this works in the storytelling example, even though $n = 5$ is certainly not a large sample.

EXAMPLE 14.10 For $n = 5$ observations, we saw in Example 14.9 that $\mu_{W^+} = 7.5$. The standard deviation of W^+ under the null hypothesis is

$$\sigma_{W^+} = \sqrt{\frac{n(n+1)(2n+1)}{24}} = \sqrt{\frac{(5)(6)(25)}{24}} = \sqrt{31.25} = 5.590$$

The continuity correction calculates the P-value $P(W^+ \geq 9)$ as $P(W^+ \geq 8.5)$, treating the value $W^+ = 9$ as occupying the interval from 8.5 to 9.5. We find the normal approximation for the P-value by standardizing and using the standard normal table:

$$P(W^+ \geq 8.5) = P\left(\frac{W^+ - 7.5}{5.590} \geq \frac{9 - 7.5}{5.590}\right) = P(Z \geq 0.27) = 0.394$$

Despite the small sample size, the normal approximation gives a result quite close to the exact value $P = 0.4062$.

14.2 The Wilcoxon Signed Rank Test

Ties

Ties among the absolute differences are handled by assigning average ranks. A tie *within* a pair creates a difference of zero. Because these are neither positive nor negative, we drop such pairs from our sample. As in the case of the Wilcoxon rank sum, ties complicate finding a *P*-value. There is no longer a usable exact distribution for the signed rank statistic W^+, and the standard deviation σ_{W^+} must be adjusted for the ties before we can use the normal approximation. Software will do this. Here is an example.

EXAMPLE 14.11

Here are the golf scores of 12 members of a college women's golf team in two rounds of tournament play. (A golf score is the number of strokes required to complete the course, so that low scores are better.)

Player	1	2	3	4	5	6	7	8	9	10	11	12
Round 2	94	85	89	89	81	76	107	89	87	91	88	80
Round 1	89	90	87	95	86	81	102	105	83	88	91	79
Difference	5	−5	2	−6	−5	−5	5	−16	4	3	−3	1

Negative differences indicate better (lower) scores on the second round. We see that 6 of the 12 golfers improved their scores. We would like to test the hypotheses that in a large population of collegiate women golfers

H_0 : scores have the same distribution in Rounds 1 and 2
H_a : scores are systematically lower or higher in Round 2

A normal quantile plot of the differences (Figure 14.8) shows some irregularity and a low outlier. We will use the Wilcoxon signed rank test.

The absolute values of the differences, with boldface indicating those that are negative, are

$$5 \ \mathbf{5} \ 2 \ \mathbf{6} \ \mathbf{5} \ \mathbf{5} \ 5 \ \mathbf{16} \ 4 \ 3 \ \mathbf{3} \ 1$$

Arrange these in increasing order and assign ranks, keeping track of which values were originally negative. Tied values receive the average of their ranks.

Absolute value	1	2	**3**	3	4	**5**	5	**5**	5	**5**	6	16
Rank	1	2	**3.5**	3.5	5	**8**	8	**8**	8	**8**	11	12

The Wilcoxon signed rank statistic is the sum of the ranks of the negative differences. (We could equally well use the sum for the ranks of the positive differences.) Its value is $W^+ = 50.5$.

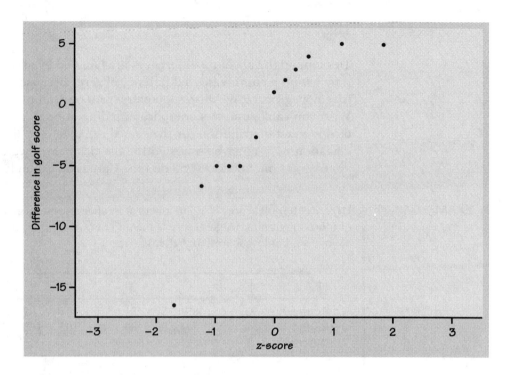

FIGURE 14.8 Normal quantile plot for Example 14.11 of the differences in scores for two rounds of a golf tournament.

EXAMPLE 14.12

Here are the two-sided P-values for the Wilcoxon signed rank test for the golf score data from several statistical programs:

Program	P-value
Data Desk	$P = 0.366$
Minitab	$P = 0.388$
SAS	$P = 0.388$
S-PLUS	$P = 0.384$

All lead to the same practical conclusion: these data give no evidence for a systematic change in scores between rounds. However, the P-values reported differ a bit from program to program. The reason for the variations is that the programs use slightly different versions of the approximate calculations needed when ties are present. The exact result depends on which of these variations the programmer chooses to use.

For these data, the matched pairs t test gives $t = 0.9314$ with $P = 0.3716$. Once again, t and W^+ lead to the same conclusion.

SUMMARY

The **Wilcoxon signed rank test** applies to matched pairs studies. It tests the null hypothesis that there is no systematic difference within pairs against alternatives that assert a systematic difference (either one-sided or two-sided).

The test is based on the **Wilcoxon signed rank statistic W^+**, which is the sum of the ranks of the positive (or negative) differences when we rank the absolute values of the differences. The **matched pairs t test** and the **sign test** are alternative tests in this setting.

***P*-values** for the signed rank test are based on the sampling distribution of W^+ when the null hypothesis is true. You can find *P*-values from special tables, software, or a normal approximation (with continuity correction).

SECTION 14.2 EXERCISES

Statistical software is very helpful in doing these exercises. If you do not have access to software, base your work on the normal approximation with continuity correction.

14.12 Table 7.1 (page 514) presents the scores on a test of understanding of spoken French for a group of high school French teachers before and after a summer language institute. The improvements in scores between the pretest and the posttest for the 20 teachers were

| 2 | 0 | 6 | 6 | 3 | 3 | 2 | 3 | −6 | 6 | 6 | 6 | 3 | 0 | 1 | 1 | 0 | 2 | 3 | 3 |

A normal quantile plot (Figure 7.7, page 515) shows granularity and a low outlier. We might wish to avoid the assumption of normality by using a rank test. Use the Wilcoxon signed rank procedure to reach a conclusion about the effect of the language institute. State hypotheses in words and report the statistic W^+, its *P*-value, and your conclusion. (Note that there are many ties in the data.)

14.13 Exercise 7.35 (page 532) gives the scores on a test of comprehension of spoken Spanish for 20 teachers before and after they attended a summer language institute. We want to know whether the institute improves Spanish comprehension.

(a) State the null and alternative hypotheses.

(b) Explain why the Wilcoxon rank sum test is *not* appropriate.

(c) Give numerical measures that describe what the data show. Then use the Wilcoxon signed rank test to assess significance. What do you conclude?

14.14 Show the assignment of ranks and the calculation of the signed rank statistic W^+ for the data in Exercise 14.12. Remember that zeros are dropped from the data before ranking, so that n is the number of nonzero differences within pairs.

14.15 Example 14.6 describes a study of the attitudes of people attending outdoor fairs about the safety of the food served at such locations. The full data set is stored on the CD as the file eg14.06.dat. It contains the responses of 303 people to several questions. The variables in this data set are (in order):

| subject | hfair | sfair | sfast | srest | gender |

The variable "sfair" contains responses to the safety question described in Example 14.6. The variable "srest" contains responses to the same question asked about food served in restaurants. We suspect that restaurant food will appear safer than food served outdoors at a fair. Do the data give good evidence for this suspicion? (Give descriptive measures, a test statistic and its P-value, and your conclusion.)

14.16 The food safety survey data described in Example 14.6 and Exercise 14.15 also contain the responses of the 303 subjects to the same question asked about food served at fast-food restaurants. These responses are the values of the variable "sfast." Is there a systematic difference between the level of concern about food safety at outdoor fairs and at fast-food restaurants?

14.17 Differences of electric potential occur naturally from point to point on a body's skin. Is the natural electric field strength best for helping wounds to heal? If so, changing the field will slow healing. The research subjects are anesthetized newts. Make a razor cut in both hind limbs. Let one heal naturally (the control). Use an electrode to change the electric field in the other to half its normal value. After two hours, measure the healing rate. Here are healing rates (in micrometers per hour) for 14 newts. (Data provided by Drina Iglesia, Purdue University. The study results are reported in D. D. S. Iglesia, E. J. Cragoe, Jr., and J. W. Vanable, "Electric field strength and epithelization in the newt (*Notophthalmus viridescens*)," *Journal of Experimental Zoology*, 274 (1996), pp. 56–62.)

Newt	Experimental limb	Control limb	Difference in healing
01	24	25	−1
02	23	13	10
03	47	44	3
04	42	45	−3
05	26	57	−31
06	46	42	4
07	38	50	−12
08	33	36	−3
09	28	35	−7
10	28	38	−10
11	21	43	−22
12	27	31	−4
13	25	26	−1
14	45	48	−3

The researchers want to know if changing the electric field reduces the healing rate for newts. State hypotheses, carry out a test, and give your conclusion. Be sure to include a description of what the data show in addition to the test results. (The researchers compared several field strengths and concluded that the natural strength is about right for fastest healing.)

14.18 Exercise 7.19 (page 527) presents these data on the vitamin C content of Haitian "bouillie" before and after cooking (in milligrams per 100 grams):

Sample	1	2	3	4	5
Before	73	79	86	88	78
After	20	27	29	36	17

Is there a significant loss of vitamin C in cooking? Use a test that does not require normality.

14.19 Exercise 7.32 (page 531) gives data on the vitamin C content of 27 bags of wheat soy blend at the factory and five months later in Haiti. We want to know if vitamin C has been lost during transportation and storage. Describe what the data show about this question. Then use a rank test to see whether there has been a significant loss.

14.20 Exercise 7.33 (page 532) contains data from a student project that investigated whether right-handed people can turn a knob faster clockwise than they can counterclockwise. Describe what the data show, then state hypotheses and do a test that does not require normality. Report your conclusions carefully.

14.3 The Kruskal-Wallis Test

We have now considered alternatives to the matched pairs and two-sample t tests for comparing the magnitude of responses to two treatments. To compare more than two treatments, we use one-way analysis of variance (ANOVA) if the distributions of the responses to each treatment are at least roughly normal and have similar spreads. What can we do when these distribution requirements are violated?

EXAMPLE 14.13 Lamb's-quarter is a common weed that interferes with the growth of corn. A researcher planted corn at the same rate in 16 small plots of ground, then randomly assigned the plots to four groups. He weeded the plots by hand to allow a fixed number of lamb's-quarter plants to grow in each meter of corn row. These numbers were 0, 1, 3, and 9 in the four groups of plots. No other weeds were allowed to grow, and all plots received identical treatment except for the weeds. Here are the yields of corn (bushels per acre) in each of the plots:[5]

Weeds per meter	Corn yield	Weeds per meter	Corn yield	Weeds per meter	Corn yield	Weeds per meter	Corn yield
0	166.7	1	166.2	3	158.6	9	162.8
0	172.2	1	157.3	3	176.4	9	142.4
0	165.0	1	166.7	3	153.1	9	162.7
0	176.9	1	161.1	3	156.0	9	162.4

The summary statistics are as follows:

Weeds	n	Mean	Std. dev.
0	4	170.200	5.422
1	4	162.825	4.469
3	4	161.025	10.493
9	4	157.575	10.118

The sample standard deviations do not satisfy our rule of thumb that for safe use of ANOVA the largest should not exceed twice the smallest. Normal quantile plots (Figure 14.9) show that outliers are present in the yields for 3 and 9 weeds per meter. These are the correct yields for their plots, so we have no justification for removing them. We may want to use a nonparametric test.

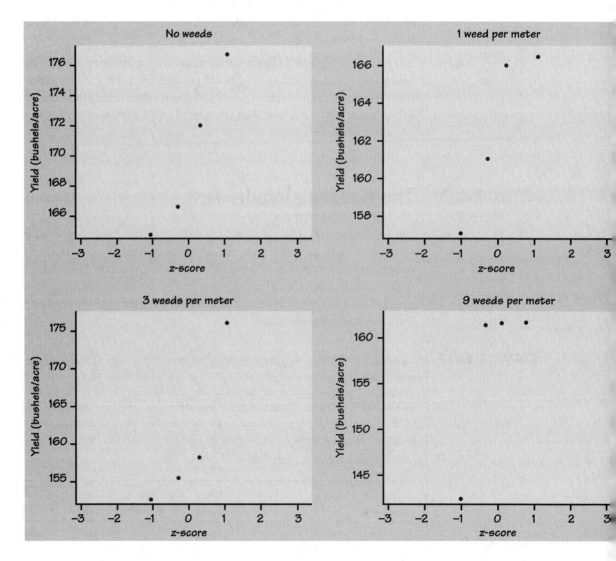

FIGURE 14.9 Normal quantile plots for the corn yields in the four treatment groups in Example 14.13.

14.3 The Kruskal-Wallis Test

Hypotheses and assumptions

The ANOVA F test concerns the means of the several populations represented by our samples. For Example 14.13, the ANOVA hypotheses are

$$H_0: \mu_0 = \mu_1 = \mu_3 = \mu_9$$
$$H_a: \text{not all four means are equal}$$

For example, μ_0 is the mean yield in the population of all corn planted under the conditions of the experiment with no weeds present. The data should consist of four independent random samples from the four populations, all normally distributed with the same standard deviation.

The *Kruskal-Wallis test* is a rank test that can replace the ANOVA F test. The assumption about data production (independent random samples from each population) remains important, but we can relax the normality assumption. We assume only that the response has a continuous distribution in each population. The hypotheses tested in our example are:

H_0: yields have the same distribution in all groups
H_a: yields are systematically higher in some groups than in others

"Systematically higher" has the precise meaning described in Section 14.1. If all of the population distributions have the same shape (normal or not), these hypotheses take a simpler form. The null hypothesis is that all four populations have the same *median* yield. The alternative hypothesis is that not all four median yields are equal.

The Kruskal-Wallis test

Recall the analysis of variance idea: we write the total observed variation in the responses as the sum of two parts, one measuring variation among the groups (sum of squares for groups, SSG) and one measuring variation among individual observations within the same group (sum of squares for error, SSE). The ANOVA F test, roughly speaking, rejects the null hypothesis that the mean responses are equal in all groups if SSG is large relative to SSE.

The idea of the Kruskal-Wallis rank test is to rank all the responses from all groups together and then apply one-way ANOVA to the ranks rather than to the original observations. If there are N observations in all, the ranks are always the whole numbers from 1 to N. The total sum of squares for the ranks is therefore a fixed number no matter what the data are. So we do not need to look at both SSG and SSE. Although it isn't obvious without some unpleasant algebra, the Kruskal-Wallis test statistic is essentially just SSG for the ranks. We give the formula, but you should rely on software to do the arithmetic. When SSG is large, that is evidence that the groups differ.

> **The Kruskal-Wallis test**
>
> Draw independent SRSs of sizes $n_1, n_2, \ldots, n_I$ from I populations. There are N observations in all. Rank all N observations and let R_i be the sum of the ranks for the ith sample. The **Kruskal-Wallis statistic** is
>
> $$H = \frac{12}{N(N+1)} \sum \frac{R_i^2}{n_i} - 3(N+1)$$
>
> When the sample sizes n_i are large and all I populations have the same continuous distribution, H has approximately the chi-square distribution with $I - 1$ degrees of freedom.
>
> The **Kruskal-Wallis test** rejects the null hypothesis that all populations have the same distribution when H is large.

We now see that, like the Wilcoxon rank sum statistic, the Kruskal-Wallis statistic is based on the sums of the ranks for the groups we are comparing. The more different these sums are, the stronger is the evidence that responses are systematically larger in some groups than in others.

The exact distribution of the Kruskal-Wallis statistic H under the null hypothesis depends on all the sample sizes n_1 to n_I, so tables are awkward. The calculation of the exact distribution is so time-consuming for all but the smallest problems that even most statistical software uses the chi-square approximation to obtain P-values. As usual, there is no usable exact distribution when there are ties among the responses. We again assign average ranks to tied observations.

EXAMPLE 14.14

In Example 14.13, there are $I = 4$ populations and $N = 16$ observations. The sample sizes are equal, $n_i = 4$. The 16 observations arranged in increasing order, with their ranks, are

Yield	142.4	153.1	156.0	157.3	158.6	161.1	162.4	162.7
Rank	1	2	3	4	5	6	7	8

Yield	162.8	165.0	166.2	166.7	166.7	172.2	176.4	176.9
Rank	9	10	11	12.5	12.5	14	15	16

There is one pair of tied observations. The ranks for each of the four treatments are

Weeds	Ranks				Sum of ranks
0	10	12.5	14	16	52.5
1	4	6	11	12.5	33.5
3	2	3	5	15	25.0
9	1	7	8	9	25.0

14.3 The Kruskal-Wallis Test

```
        Wilcoxon Scores (Rank Sums) for Variable YIELD
                 Classified by Variable WEEDS

                    Sum of      Expected     Std Dev        Mean
WEEDS       N       Scores      Under H0     Under H0       Score

0           4      52.5000000     34.0      8.24014563    13.1250000
1           4      33.5000000     34.0      8.24014563     8.3750000
3           4      25.0000000     34.0      8.24014563     6.2500000
9           4      25.0000000     34.0      8.24014563     6.2500000
              Average Scores Were Used for Ties


Kruskal-Wallis Test (Chi-Square Approximation)
 CHISQ =   5.5725   DF =   3   Prob > CHISQ = 0.1344
```

FIGURE 14.10 Output from SAS for the Kruskal-Wallis test applied to the data in Example 14.13. SAS uses the chi-square approximation to obtain a P-value.

The Kruskal-Wallis statistic is therefore

$$H = \frac{12}{N(N+1)} \sum \frac{R_i^2}{n_i} - 3(N+1)$$

$$= \frac{12}{(16)(17)} \left(\frac{52.5^2}{4} + \frac{33.5^2}{4} + \frac{25^2}{4} + \frac{25^2}{4} \right) - (3)(17)$$

$$= \frac{12}{272}(1282.125) - 51 = 5.56$$

Referring to the table of chi-square critical points (Table G) with df = 3, we find that the P-value lies in the interval $0.10 < P < 0.15$. This small experiment suggests that more weeds decrease yield but does not provide convincing evidence that weeds have an effect.

Figure 14.10 displays the output from the SAS statistical software, which gives the results $H = 5.5725$ and $P = 0.1344$. The software makes a small adjustment for the presence of ties that accounts for the slightly larger value of H. The adjustment makes the chi-square approximation more accurate. It would be important if there were many ties.

As an option, SAS will calculate the exact P-value for the Kruskal-Wallis test. The result for Example 14.14 is $P = 0.1299$. This result required several hours of computing time.[6] Fortunately, the chi-square approximation is quite accurate. The ordinary ANOVA F test gives $F = 1.73$ with $P = 0.2130$. Although the practical conclusion is the same, ANOVA and Kruskal-Wallis do not agree closely in this example. The rank test is more reliable for these small samples with outliers.

SUMMARY

The **Kruskal-Wallis test** compares several populations on the basis of independent random samples from each population. This is the **one-way analysis of variance** setting.

The null hypothesis for the Kruskal-Wallis test is that the distribution of the response variable is the same in all the populations. The alternative hypothesis is that responses are systematically larger in some populations than in others.

The **Kruskal-Wallis statistic H** can be viewed in two ways. It is essentially the result of applying one-way ANOVA to the ranks of the observations. It is also a comparison of the sums of the ranks for the several samples.

When the sample sizes are not too small and the null hypothesis is true, H for comparing I populations has approximately the chi-square distribution with $I-1$ degrees of freedom. We use this approximate distribution to obtain P-values.

SECTION 14.3 EXERCISES

Statistical software is needed to do these exercises without unpleasant hand calculations. If you do not have access to software, omit normal quantile plots, find the Kruskal-Wallis statistic H by hand, and use the chi-square table to get approximate P-values.

14.21 How do nematodes (microscopic worms) affect plant growth? A botanist prepares 16 identical planting pots and then introduces different numbers of nematodes into the pots. A tomato seedling is transplanted into each plot. Here are data on the increase in height of the seedlings (in centimeters) 16 days after planting. (Data provided by Matthew Moore.)

Nematodes	Seedling growth			
0	10.8	9.1	13.5	9.2
1,000	11.1	11.1	8.2	11.3
5,000	5.4	4.6	7.4	5.0
10,000	5.8	5.3	3.2	7.5

We applied ANOVA to these data in Exercise 12.10 (page 782). Because the samples are very small, it is difficult to assess normality.

(a) What hypotheses does ANOVA test? What hypotheses does Kruskal-Wallis test?

(b) Find the median growth in each group. Do nematodes appear to retard growth? Apply the Kruskal-Wallis test. What do you conclude?

14.22 The presence of harmful insects in farm fields is detected by erecting boards covered with a sticky material and then examining the insects trapped on the board. To investigate which colors are most attractive to cereal leaf beetles, researchers placed six boards of each of four colors in a field of oats in July. The table below gives data on the number of cereal leaf beetles

Section 14.3 Exercises

trapped. (Based on M. C. Wilson and R. E. Shade, "Relative attractiveness of various luminescent colors to the cereal leaf beetle and the meadow spittlebug," *Journal of Economic Entomology,* 60 (1967), pp. 578–580.)

Color	Insects trapped					
Lemon yellow	45	59	48	46	38	47
White	21	12	14	17	13	17
Green	37	32	15	25	39	41
Blue	16	11	20	21	14	7

We applied ANOVA to these data in Exercise 12.11 (page 782).

(a) Make a normal quantile plot for each group. Are there indications of lack of normality?

(b) What hypotheses does ANOVA test? What hypotheses does Kruskal-Wallis test?

(c) Find the median number of beetles trapped by boards of each color. Which colors appear more effective? Use the Kruskal-Wallis test to see if there are significant differences among the colors. What do you conclude?

14.23 Table 1.8 (page 40) presents data on the calorie and sodium content of selected brands of beef, meat, and poultry hot dogs. We will regard these brands as random samples from all brands available in food stores. We saw that the distribution of calories in meat hot dogs had two clusters and a low outlier. We might therefore prefer to use a nonparametric test. Give the five-number summaries for the three types of hot dog and then apply the Kruskal-Wallis test. Report your conclusions carefully.

14.24 Exercise 14.22 gives data on the counts of insects attracted by boards of four different colors. Carry out the Kruskal-Wallis test by hand, following these steps.

(a) What are I, the n_i, and N in this example?

(b) Arrange the counts in order and assign ranks. Be careful about ties. Find the sum of the ranks for each color.

(c) Calculate the Kruskal-Wallis statistic H. How many degrees of freedom should you use for the chi-square approximation of its null-hypothesis distribution? Use the chi-square table to give an approximate P-value.

14.25 Repeat the analysis of Exercise 14.23 for the sodium content of hot dogs, using the data in Table 1.8 (page 40).

14.26 Table 12.4 (page 781) gives data on the effect of four treatments on the spatial-temporal reasoning ability of preschool children. The treatments are piano lessons, singing lessons, computer instruction, and no lessons of any kind. The response variable is the change in a child's score on a test of spatial-temporal reasoning.

(a) Give the five-number summary for each group. What do the data suggest about the effects of the treatments?

(b) Make a normal quantile plot for each group. Which group deviates most from normality?

(c) Do the treatments differ significantly in their ability to improve children's spatial-temporal reasoning?

14.27 Example 14.6 describes a study of the attitudes of people attending outdoor fairs about the safety of the food served at such locations. The full data set is stored on the CD as the file eg14.06.dat. It contains the responses of 303 people to several questions. The variables in this data set are (in order):

subject	hfair	sfair	sfast	srest	gender

The variable "sfair" contains responses to the safety question described in Example 14.6. The variables "srest" and "sfast" contain responses to the same question asked about food served in restaurants and in fast-food chains. Explain carefully why we *cannot* use the Kruskal-Wallis test to see if there are systematic differences in perceptions of food safety in these three locations.

14.28 In a study of heart disease in male federal employees, researchers classified 356 volunteer subjects according to their socioeconomic status (SES) and their smoking habits. There were three categories of SES: high, middle, and low. Individuals were asked whether they were current smokers, former smokers, or had never smoked. (Ray H. Rosenman et al., "A 4-year prospective study of the relationship of different habitual vocational physical activity to risk and incidence of ischemic heart disease in volunteer male federal employees," in P. Milvey (ed.), *The Marathon: Physiological, Medical, Epidemiological and Psychological Studies*, New York Academy of Sciences, 301 (1977), pp. 627–641.) Here are the data, as a two-way table of counts:

SES	Never (1)	Former (2)	Current (3)
High	68	92	51
Middle	9	21	22
Low	22	28	43

The data for all 356 subjects are stored in the file ex14.28.dat on the CD. Smoking behavior is stored numerically as 1, 2, or 3 using the codes given in the column headings above.

(a) Higher SES people in the United States smoke less as a group than lower SES people. Do these data show a relationship of this kind? Give percents that back your statements.

(b) Apply the chi-square test to see if there is a significant relationship between SES and smoking behavior.

(c) The chi-square test ignores the ordering of the responses. Use the Kruskal-Wallis test (with many ties) to test the hypothesis that some SES classes smoke systematically more than others.

NOTES

14.29 (Optional)
multiple comparisons

As in ANOVA, we often want to carry out a **multiple comparisons** procedure following a Kruskal-Wallis test to tell us *which* groups differ significantly.[7] Here is a simple method: If we carry out k tests at fixed significance level $0.05/k$, the probability of *any* false rejection among the k tests is always no greater than 0.05. That is, to get overall significance level 0.05 for all of k comparisons, do each individual comparison at the $0.05/k$ level. In Exercise 14.23 you found a significant difference among the calorie contents of three types of hot dog. Now we will explore multiple comparisons.

(a) Write down all of the pairwise comparisons we can make, for example, beef versus meat. There are three possible pairwise comparisons.

(b) Carry out three Wilcoxon rank sum tests, one for each of the three pairs of hot dog types. What are the three two-sided P-values?

(c) For purposes of multiple comparisons, any of these three tests is significant if its P-value is no greater than $0.05/3 = 0.0167$. Which pairs differ significantly at the overall 0.05 level?

14.30 (Optional)

Exercise 14.29 outlines how to use the Wilcoxon rank sum test several times for multiple comparisons with overall significance level 0.05 for all comparisons together. In Exercise 14.22 you found that the numbers of beetles attracted by boards of four colors differ significantly. At the overall 0.05 level, which pairs of colors differ significantly? (Hint: There are 6 possible pairwise comparisons among 4 colors.)

NOTES

1. Data provided by Sam Phillips, Purdue University. The data have been slightly modified to remove one tie to simplify Exercise 14.4.

2. For purists, here is the precise definition: X_1 is *stochastically larger* than X_2 if

$$P(X_1 > a) \geq P(X_2 > a)$$

for all a, with strict inequality for at least one a. The Wilcoxon rank sum test is effective against this alternative in the sense that the power of the test approaches 1 (that is, the test becomes more certain to reject the null hypothesis) as the number of observations increases.

3. Data from Huey Chern Boo, "Consumers' perceptions and concerns about safety and healthfulness of food served at fairs and festivals," M.S. thesis, Purdue University, 1997.

4. Data provided by Susan Stadler, Purdue University.

5. See Note 1.

6. Using SAS Version 6.12 on a 166 MHz Pentium personal computer (a quite fast machine at the time this was written) required about 3.5 hours to obtain the exact P-value in Example 14.14. No wonder very few software systems offer exact P-values for the Kruskal-Wallis statistic.

7. For more details on multiple comparisons (but not the simple procedure given here), see M. Hollander and D. A. Wolfe, *Nonparametric Statistical Methods*, Wiley, New York, 1973. This book is a useful reference on applied aspects of nonparametric inference in general.

Prelude

The simple and multiple linear regression methods we studied in Chapters 10 and 11 are used to model the relationship between a quantitative response variable and one or more explanatory variables. A key assumption for these models is that the deviations from the model fit are normally distributed. In this chapter we describe similar methods that are used when the response variable has only two possible values.

- How does the concentration of an insecticide relate to whether or not an insect is killed?
- To what extent does gender predict whether or not a college student will be a binge drinker?
- Is high blood pressure associated with an increased risk of death from cardiovascular disease?

CHAPTER 15
Logistic Regression

Our response variable has only two values: success or failure, live or die, acceptable or not. If we let the two values be 1 and 0, the mean is the proportion of ones, $p = P(\text{success})$. With n independent observations, we have the *binomial setting* (page 376). What is *new* here is that we have data on an *explanatory variable* x. We study how p depends on x. For example, suppose we are studying whether a patient lives ($y = 1$) or dies ($y = 0$) after being admitted to a hospital. Here, p is the probability that a patient lives, and possible explanatory variables include (a) whether the patient is in good condition or in poor condition, (b) the type of medical problem that the patient has, and (c) the age of the patient. Note that the explanatory variables can be either categorical or quantitative. Logistic regression[1] is a statistical method for describing these kinds of relationships.

Binomial distributions and odds

In Chapter 5 we studied binomial distributions and in Chapter 8 we learned how to do statistical inference for the proportion p of successes in the binomial setting. We start with a brief review of some of these ideas that we will need in this chapter.

EXAMPLE 15.1

Example 8.1 describes a survey of 17,096 students in U.S. four-year colleges. The researchers were interested in estimating the proportion of students who are frequent binge drinkers. A student who reports drinking five or more drinks in a row three or more times in the past two weeks is called a frequent binge drinker. In the notation of Chapter 5, p is the proportion of frequent binge drinkers in the entire population of college students in U.S. four-year colleges. The number of frequent binge drinkers in an SRS of size n has the binomial distribution with parameters n and p. The sample size is $n = 17,096$ and the number of frequent binge drinkers in the sample is 3314. The sample proportion is

$$\hat{p} = \frac{3314}{17,096} = 0.1938$$

odds

Logistic regressions work with **odds** rather than proportions. The odds are simply the ratio of the proportions for the two possible outcomes. If $\hat{p}$ is the proportion for one outcome, then $1 - \hat{p}$ is the proportion for the second outcome.

$$\text{ODDS} = \frac{\hat{p}}{1 - \hat{p}}$$

A similar formula for the population odds is obtained by substituting p for $\hat{p}$ in this expression.

EXAMPLE 15.2

For the binge-drinking data the proportion of frequent binge drinkers in the sample is $\hat{p} = 0.1938$, so the proportion of students who are not frequent binge drinkers is

$$1 - \hat{p} = 1 - 0.1938 = 0.8062$$

Therefore, the odds of a student being a frequent binge drinker are

$$\text{ODDS} = \frac{\hat{p}}{1-\hat{p}}$$
$$= \frac{0.1938}{0.8062}$$
$$= 0.24$$

When people speak about odds, they often round to integers or fractions. Since 0.24 is approximately 1/4, we could say that the odds that a college student is a frequent binge drinker are 1 to 4. In a similar way, we could describe the odds that a college student is *not* a frequent binge drinker as 4 to 1.

In Example 8.8 (page 603) we compared the proportions of frequent binge drinkers among men and women college students using a confidence interval. There we found that the proportion for men was 0.227 (22.7%) and that the proportion for women was 0.170 (17.0%). The difference is 0.057 and the 95% confidence interval is (0.045, 0.069). We can summarize this result by saying, "The proportion of frequent binge drinkers is 5.7% higher among men than among women."

Another way to analyze these data is to use logistic regression. The explanatory variable is gender, a categorical variable. To use this in a regression (logistic or otherwise), we need to use a numeric code. The usual way to do this is with an **indicator variable**. For our problem we will use an indicator of whether or not the student is a man:

indicator variable

$$x = \begin{cases} 1 & \text{if the student is a man} \\ 0 & \text{if the student is a woman} \end{cases}$$

The response variable is the proportion of frequent binge drinkers. For use in a logistic regression, we perform two transformations on this variable. First, we convert to odds. For men,

$$\text{ODDS} = \frac{\hat{p}}{1-\hat{p}}$$
$$= \frac{0.227}{1-0.227}$$
$$= 0.294$$

Similarly, for women we have

$$\text{ODDS} = \frac{\hat{p}}{1-\hat{p}}$$
$$= \frac{0.170}{1-0.170}$$
$$= 0.205$$

The logistic regression model

In simple linear regression we modeled the mean μ of the response variable y as a linear function of the explanatory variable: $\mu = \beta_0 + \beta_1 x$. With logistic regression we are interested in modeling the mean of the response variable p in terms of an explanatory variable x. We could try to relate p and x through the equation $p = \beta_0 + \beta_1 x$. Unfortunately, this is not a good model. As long as $\beta_1 \neq 0$, extreme values of x will give values of $\beta_0 + \beta_1 x$ that are inconsistent with the fact that $0 \leq p \leq 1$.

The logistic regression solution to this difficulty is to transform the odds $(p/(1-p))$ using the natural logarithm. We use the term *log odds* for this transformation. We model the log odds as a linear function of the explanatory variable:

$$\log\left(\frac{p}{1-p}\right) = \beta_0 + \beta_1 x$$

Figure 15.1 graphs the relationship between p and x for some different values of β_0 and β_1. For logistic regression we use *natural* logarithms. There are tables of natural logarithms, and many calculators have a built-in function for this transformation. As we did with linear regression, we use y for the response variable. So for men,

$$y = \log(\text{ODDS}) = \log(0.294) = -1.23$$

and for women,

$$y = \log(\text{ODDS}) = \log(0.205) = -1.59$$

In these expressions we use y as the observed value of the response variable, the log odds of being a frequent binge drinker. We are now ready to build the logistic regression model.

> **Logistic Regression Model**
>
> The **statistical model for logistic regression** is
>
> $$\log\left(\frac{p}{1-p}\right) = \beta_0 + \beta_1 x$$
>
> where the p is a binomial proportion and x is the explanatory variable. The parameters of the logistic model are β_0 and β_1.

EXAMPLE 15.3 For our binge-drinking example, there are $n = 17{,}096$ students in the sample. The explanatory variable is gender, which we have coded using an indicator variable with values $x = 1$ for men and $x = 0$ for women. The response variable is also an indicator variable. Thus, the student is either a frequent binge drinker or the student is not a frequent binge drinker. Think of the process of randomly selecting a student and recording the values of x and whether or not the student is a frequent binge drinker. The model says that the probability (p) that this student is a frequent binge drinker depends upon the student's gender ($x = 1$ or $x = 0$). So there are two possible values for p, say p_{men} and p_{women}.

Logistic regression with an indicator explanatory variable is a very special case. It is important because many multiple logistic regression analyses focus on one or more such variables as the primary explanatory variables of interest. For now, we use this special case to understand a little more about the model.

Chapter 15: Logistic Regression

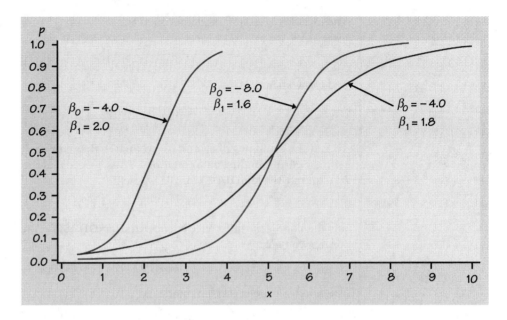

FIGURE 15.1 Plot of p versus x for selected values of β_0 and β_1.

The logistic regression model specifies the relationship between p and x. Since there are only two values for x, we write both equations. For men,

$$\log\left(\frac{p_{\text{men}}}{1 - p_{\text{men}}}\right) = \beta_0 + \beta_1$$

and for women

$$\log\left(\frac{p_{\text{women}}}{1 - p_{\text{women}}}\right) = \beta_0$$

Note that there is a β_1 term in the equation for men because $x = 1$ but it is missing in the equation for women because $x = 0$.

Fitting and interpreting the logistic regression model

In general the calculations needed to find estimates b_0 and b_1 for the parameters β_0 and β_1 are complex and require the use of software. When the explanatory variable has only two possible values, however, we can easily find the estimates. This simple framework also provides a setting where we can learn what the logistic regression parameters mean.

EXAMPLE 15.4 In the binge-drinking example, we found the log odds for men

$$y = \log\left(\frac{\hat{p}_{\text{men}}}{1 - \hat{p}_{\text{men}}}\right) = -1.23$$

and for women

$$y = \log\left(\frac{\hat{p}_{\text{women}}}{1 - \hat{p}_{\text{women}}}\right) = -1.59$$

The logistic regression model for men is

$$\log\left(\frac{p_{\text{men}}}{1-p_{\text{men}}}\right) = \beta_0 + \beta_1$$

and for women, it is

$$\log\left(\frac{p_{\text{women}}}{1-p_{\text{women}}}\right) = \beta_0$$

To find the estimates of b_0 and b_1, we match the male and female model equations with the corresponding data equations. Thus, we see that the estimate of the intercept b_0 is simply the log(ODDS) for the women:

$$b_0 = -1.59$$

and the slope is the difference between the log(ODDS) for the men and the log(ODDS) for the women:

$$b_1 = -1.23 - (-1.59) = 0.36$$

The fitted logistic regression model is

$$\log(\text{ODDS}) = -1.59 + 0.36x$$

The slope in this logistic regression model is the difference between the log(ODDS) for men and the log(ODDS) for women. Most people are not comfortable thinking in the log(ODDS) scale, so interpretation of the results in terms of the regression slope is difficult. Usually, we apply a transformation to help us. With a little algebra, it can be shown that

$$\frac{\text{ODDS}_{\text{men}}}{\text{ODDS}_{\text{women}}} = e^{0.36} = 1.43$$

odds ratio

The transformation $e^{0.36}$ undoes the logarithm and transforms the logistic regression slope into an **odds ratio**, in this case, the ratio of the odds that a man is a frequent binge drinker to the odds that a woman is a frequent binge drinker. In other words, we can multiply the odds for women by the odds ratio to obtain the odds for men:

$$\text{ODDS}_{\text{men}} = 1.43 \times \text{ODDS}_{\text{women}}$$

In this case, the odds for men are 1.43 times the odds for women.

Notice that we have chosen the coding for the indicator variable so that the regression slope is positive. This will give an odds ratio that is greater than 1. Had we coded women as 1 and men as 0, the signs of the parameters would be reversed, the fitted equation would be $\log(\text{ODDS}) = 1.59 - 0.36x$, and the odds ratio would be $e^{-0.036} = 0.70$. The odds for women are 70% of the odds for men.

Logistic regression with an explanatory variable having two values is a very important special case. Here is an example where the explanatory variable is quantitative.

EXAMPLE 15.5 | The CHEESE data set described in the Data Appendix includes a response variable called "Taste" that is a measure of the quality of the cheese obtained from several tasters. For this example, we will classify the cheese as acceptable (tasteok = 1) if

Taste $\geq$ 37 and unacceptable (tasteok = 0) if Taste < 37. This is our response variable. The data set contains three explanatory variables: "Acetic," "H2S," and "Lactic." Let's use Acetic as the explanatory variable. The model is

$$\log\left(\frac{p}{1-p}\right) = \beta_0 + \beta_1 x$$

where p is the probability that the cheese is acceptable and x is the value of Acetic. The model for estimated log odds fitted by software is

$$\log(\text{ODDS}) = b_0 + b_1 x = -13.71 + 2.25x$$

The odds ratio is $e^{b_1} = 9.48$. This means that if we increase the acetic acid content x by one unit, we increase the odds that the cheese will be acceptable by about 9.5 times.

Inference for logistic regression

Statistical inference for logistic regression is very similar to statistical inference for simple linear regression. We calculate estimates of the model parameters and standard errors for these estimates. Confidence intervals are formed in the usual way, but we use standard normal z^*-values rather than critical values from the t distributions. The ratio of the estimate to the standard error is the basis for hypothesis tests. Often the test statistics are given as the squares of these ratios, and in this case the P-values are obtained from the chi-square distributions with 1 degree of freedom.

Confidence Intervals and Significance Tests for Logistic Regression Parameters

A **level C confidence interval for the slope β_1** is

$$b_1 \pm z^* \text{SE}_{b_1}$$

The ratio of the odds for a value of the explanatory variable equal to $x + 1$ to the odds for a value of the explanatory variable equal to x is the **odds ratio**.

A **level C confidence interval for the odds ratio e^{β_1}** is obtained by transforming the confidence interval for the slope:

$$(e^{b_1 - z^* \text{SE}_{b_1}}, \; e^{b_1 + z^* \text{SE}_{b_1}})$$

In these expressions z^* is the value for the standard normal density curve with area C between $-z^*$ and z^*.

To test the hypothesis $H_0: \beta_1 = 0$, compute the **test statistic**

$$X^2 = \left(\frac{b_1}{\text{SE}_{b_1}}\right)^2$$

In terms of a random variable X^2 having approximately a χ^2 distribution with 1 degree of freedom, the P-value for a test of H_0 against $H_a: \beta_1 \neq 0$ is $P(\chi^2 \geq X^2)$.

Variable	DF	Parameter Estimate	Standard Error	Wald Chi-Square	Pr > Chi-Square	Odds Ratio
INTERCPT	1	-1.5869	0.0267	3520.4040	0.0001	.
X	1	0.3616	0.0388	86.6714	0.0001	1.436

FIGURE 15.2 Logistic regression output for the binge-drinking data, for Example 15.6.

We have expressed the hypothesis-testing framework in terms of the slope β_1 because this form closely resembles what we studied in simple linear regression. In many applications, however, the results are expressed in terms of the odds ratio. A slope of 0 is the same as an odds ratio of 1, so we often express the null hypothesis of interest as "the odds ratio is 1." This means that the two odds are equal and the explanatory variable is not useful for predicting the odds.

EXAMPLE 15.6 Figure 15.2 gives the output from the SAS logistic procedure for the binge-drinking example. The parameter estimates are given as $b_0 = -1.5869$ and $b_1 = 0.3616$, the same as we calculated directly in Example 15.4, but with more significant digits. The standard errors are 0.0267 and 0.0388. A 95% confidence interval for the slope is

$$b_1 \pm z^* SE_{b_1} = 0.3616 \pm (1.96)(0.0388)$$
$$= 0.3616 \pm 0.0760$$

We are 95% confident that the slope is between 0.2855 and 0.4376. The output provides the odds ratio 1.436 but does not give the confidence interval. This is easy to compute from the interval for the slope:

$$(e^{b_1 - z^* SE_{b_1}}, e^{b_1 + z^* SE_{b_1}}) = (e^{0.2855}, e^{0.4376})$$
$$= (1.33, 1.55)$$

For this problem we would report, "College men are more likely to be frequent binge drinkers than college women (odds ratio = 1.44, 95% CI is 1.33 to 1.55)."

In applications such as these, it is standard to use 95% for the confidence coefficient. With this convention, the confidence interval gives us the result of testing the null hypothesis that the odds ratio is 1 for a significance level of 0.05. If the confidence interval does not include 1, we reject H_0 and conclude that the odds for the two groups are different; if not, the data do not provide enough evidence to distinguish the groups in this way.

The following example is typical of many applications of logistic regression. Here there is a designed experiment with five different values for the explanatory variable.

EXAMPLE 15.7 An experiment was designed to examine how well the insecticide rotenone kills aphids that feed on the chrysanthemum plant called *Macrosiphoniella sanborni*.[2] The explanatory variable is the log concentration (in milligrams per liter) of the insecticide. At each concentration, approximately 50 insects were exposed. Each insect was either killed or not killed. We summarize the data using the number killed.

Chapter 15: Logistic Regression

The response variable for logistic regression is the log odds of the proportion killed. Here are the data:

Concentration (log)	Number of insects	Number killed
0.96	50	6
1.33	48	16
1.63	46	24
2.04	49	42
2.32	50	44

If we transform the response variable (by taking log odds) and use least-squares, we get the fit illustrated in Figure 15.3. The logistic regression fit is given in Figure 15.4. It is a transformed version of Figure 15.3 with the fit calculated using the logistic model.

One of the major themes of this text is that we should present the results of a statistical analysis with a graph. For the insecticide example we have done this with Figure 15.4 and the results appear to be convincing. But suppose that rotenone has no ability to kill *Macrosiphoniella sanborni*. What is the chance that we would observe experimental results at least as convincing as what we observed if this supposition were true? The answer is the P-value for the test of the null hypothesis that the logistic regression slope is zero. If this P-value

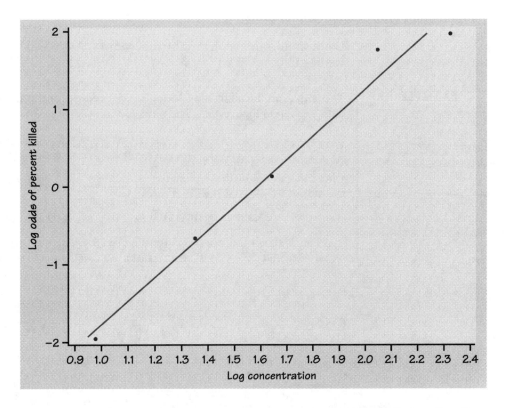

FIGURE 15.3 Plot of log odds of percent killed versus log concentration for the insecticide data, for Example 15.7.

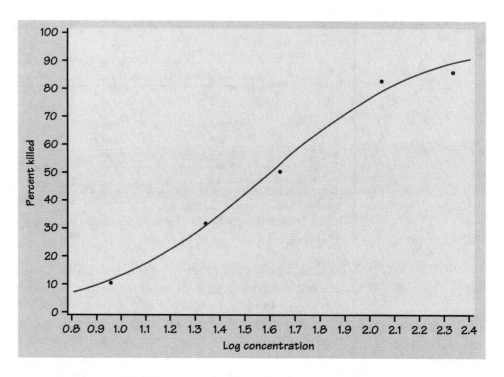

FIGURE 15.4 Plot of percent killed versus log concentration with the logistic fit for the insecticide data, for Example 15.7.

is not small, our graph may be misleading. Statistical inference provides what we need.

EXAMPLE 15.8 The output produced by the SAS logistic procedure for the analysis of the insecticide data is given in Figure 15.5. The model is

$$\log\left(\frac{p}{1-p}\right) = \beta_0 + \beta_1 x$$

where the values of the explanatory variable x are 0.96, 1.33, 1.63, 2.04, 2.32. From the output we see that the fitted model is

$$\log(\text{ODDS}) = b_0 + b_1 x = -4.89 + 3.10x$$

This is the fit that we plotted in Figure 15.4. The null hypothesis that $\beta_1 = 0$ is clearly rejected ($X^2 = 64.07$, $P < 0.001$). We calculate a 95% confidence interval for

Variable	DF	Parameter Estimate	Standard Error	Wald Chi-Square	Pr > Chi-Square	Odds Ratio
INTERCPT	1	−4.8869	0.6429	57.7757	0.0001	.
LCONC	1	3.1035	0.3877	64.0744	0.0001	22.277

FIGURE 15.5 Logistic regression output for the insecticide data, for Example 15.8.

Chapter 15: Logistic Regression

β_1 using the estimate $b_1 = 3.1035$ and its standard error $\text{SE}_{b_1} = 0.3877$ given in the output:

$$b_1 \pm z^* \text{SE}_{b_1} = 3.1035 \pm (1.96)(0.3877)$$
$$= 3.1035 \pm 0.7599$$

We are 95% confident that the true value of the slope is between 2.34 and 3.86.

The odds ratio is given on the output as 22.277. An increase of one unit in the log concentration of insecticide (x) is associated with a 22-fold increase in the odds that an insect will be killed. The confidence interval for the odds is obtained from the interval for the slope:

$$(e^{b_1 - z^* \text{SE}_{b_1}}, e^{b_1 + z^* \text{SE}_{b_1}}) = (e^{2.34361}, e^{3.86339})$$
$$= (10.42, 47.63)$$

Note again that the test of the null hypothesis that the slope is zero is the same as the test of the null hypothesis that the odds are 1. If we were reporting the results in terms of the odds, we could say, "The odds of killing an insect increase by a factor of 22.3 for each unit increase in the log concentration of insecticide ($X^2 = 64.07$, $P < 0.001$; 95% CI is 10.4 to 47.6)."

In Example 15.5 we studied the problem of predicting whether or not the taste of cheese was acceptable using Acetic as the explanatory variable. We now revisit this example and show how statistical inference is an important part of the conclusion.

EXAMPLE 15.9 The output for a logistic regression analysis using Acetic as the explanatory variable is given in Figure 15.6. In Example 15.5 we gave the fitted model:

$$\log(\text{ODDS}) = b_0 + b_1 x = -13.71 + 2.25x$$

From the output we see that because $P = 0.0285$, we can reject the null hypothesis that $\beta_1 = 0$. The value of the test statistic is $X^2 = 4.79$ with 1 degree of freedom. We use the estimate $b_1 = 2.2490$ and its standard error $\text{SE}_{b_1} = 1.0271$ to compute the 95% confidence interval for β_1:

$$b_1 \pm z^* \text{SE}_{b_1} = 2.2490 \pm (1.96)(1.0271)$$
$$= 2.2490 \pm 2.0131$$

Our estimate of the slope is 2.25 and we are 95% confident that the true value is between 0.24 and 4.26. For the odds ratio, the estimate on the output is 9.48. The 95% confidence interval is

$$(e^{b_1 - z^* \text{SE}_{b_1}}, e^{b_1 + z^* \text{SE}_{b_1}}) = (e^{0.23588}, e^{4.26212})$$
$$= (1.27, 70.96)$$

Variable	DF	Parameter Estimate	Standard Error	Wald Chi-Square	Pr > Chi-Square	Odds Ratio
INTERCPT	1	-13.7052	5.9319	5.3380	0.0209	.
ACETIC	1	2.2490	1.0271	4.7947	0.0285	9.479

FIGURE 15.6 Logistic regression output for the cheese data with Acetic as the explanatory variable, for Example 15.9.

We estimate that increasing the acetic acid content of the cheese by one unit will increase the odds that the cheese will be acceptable by about 9 times. The data, however, do not give us a very accurate estimate. The odds ratio could be as small as a little more than 1 or as large as 71 with 95% confidence. We have evidence to conclude that cheeses with higher concentrations of acetic acid are more likely to be acceptable, but establishing the true relationship accurately would require more data.

Multiple logistic regression

The cheese example that we just considered naturally leads us to the next topic. The data set includes three variables: Acetic, H2S, and Lactic. We examined the model where Acetic was used to predict the odds that the cheese was acceptable. Do the other explanatory variables contain additional information that will give us a better prediction? We use **multiple logistic regression** to answer this question. Generating the computer output is easy, just as it was when we generalized simple linear regression with one explanatory variable to multiple linear regression with more than one explanatory variable in Chapter 11. The statistical concepts are similar although the computations are more complex. Here is the example.

EXAMPLE 15.10 As in Example 15.8, we predict the odds that the cheese is acceptable. The explanatory variables are Acetic, H2S, and Lactic. Figure 15.7 gives the output. The fitted model is

$$\log(\text{ODDS}) = b_0 + b_1 \text{Acetic} + b_2 \text{H2S} + b_3 \text{Lactic}$$
$$= -14.26 + 0.58 \text{Acetic} + 0.68 \text{H2S} + 3.47 \text{Lactic}$$

When analyzing data using multiple regression, we first examine the hypothesis that all of the regression coefficients for the explanatory variables are zero. We do the same for logistic regression. The hypothesis

$$H_0: \beta_1 = \beta_2 = \beta_3 = 0$$

```
                          Intercept
               Intercept     and
Criterion        Only     Covariates     Chi-Square for Covariates
-2 LOG L        34.795      18.461          16.334 with 3 DF (p=0.0010)

              Parameter  Standard    Wald        Pr >       Odds
Variable  DF  Estimate    Error   Chi-Square  Chi-Square   Ratio

INTERCPT  1   -14.2604    8.2869    2.9613      0.0853       .
ACETIC    1     0.5845    1.5442    0.1433      0.7051      1.794
H2S       1     0.6849    0.4040    2.8730      0.0901      1.983
LACTIC    1     3.4684    2.6497    1.7135      0.1905     32.086
```

FIGURE 15.7 Logistic regression output for the cheese data with Acetic, H2S, and Lactic as the explanatory variables, for Example 15.10.

is tested by a chi-square statistic with 3 degrees of freedom. This is given in the output on the line for the criterion "−2 LOG L" under the heading "Chi-Square for Covariates." The statistic is $X^2 = 16.33$ and the P-value is 0.001. We reject H_0 and conclude that one or more of the explanatory variables can be used to predict the odds that the cheese is acceptable. We now examine the coefficients for each variable and the tests that each of these is 0. The P-values are 0.71, 0.09, and 0.19. None of the null hypotheses, H_0: $\beta_1 = 0$, H_0: $\beta_2 = 0$, and H_0: $\beta_3 = 0$, can be rejected.

Our initial multiple logistic regression analysis told us that the explanatory variables contain information that is useful for predicting whether or not the cheese is acceptable. Because the explanatory variables are correlated, however, we cannot clearly distinguish which variables or combinations of variables are important. Further analysis of these data using subsets of the three explanatory variables is needed to clarify the situation. We leave this work for the exercises.

SUMMARY

If $\hat{p}$ is the sample proportion, then the **odds** are $\hat{p}/(1-\hat{p})$, the ratio of the proportion of times the event happens to the proportion of times the event does not happen.

The **logistic regression model** relates the log of the odds to the explanatory variable:

$$\log\left(\frac{p_i}{1-p_i}\right) = \beta_0 + \beta_1 x_i$$

where the response variables for $i = 1, 2, \ldots, n$ are independent binomial random variables with parameters 1 and p_i; that is, they are independent with distributions $B(1, p_i)$. The explanatory variable is x.

The **parameters** of the logistic model are β_0 and β_1.

The **odds ratio** is e^{β_1}, where β_1 is the slope in the logistic regression model.

A **level C confidence interval for the intercept** β_0 is

$$b_0 \pm z^* \text{SE}_{b_0}$$

A **level C confidence interval for the slope** β_1 is

$$b_1 \pm z^* \text{SE}_{b_1}$$

A **level C confidence interval for the odds ratio** e^{β_1} is obtained by transforming the confidence interval for the slope:

$$(e^{b_1 - z^* \text{SE}_{b_1}}, e^{b_1 + z^* \text{SE}_{b_1}})$$

In these expressions z^* is the value for the standard normal density curve with area C between $-z^*$ and z^*.

To test the hypothesis $H_0: \beta_1 = 0$, compute the **test statistic**

$$X^2 = \left(\frac{b_1}{SE_{b_1}}\right)^2$$

In terms of a random variable X^2 having a χ^2 distribution with 1 degree of freedom, the P-value for a test of H_0 against $H_a: \beta_1 \neq 0$ is $P(\chi^2 \geq X^2)$. This is the same as testing the null hypothesis that the odds ratio is 1.

In **multiple logistic regression** the response variable has two possible values, as in logistic regression, but there can be several explanatory variables.

CHAPTER 15 EXERCISES

15.1 There is much evidence that high blood pressure is associated with increased risk of death from cardiovascular disease. A major study of this association examined 3338 men with high blood pressure and 2676 men with low blood pressure. During the period of the study, 21 men in the low-blood-pressure and 55 in the high-blood-pressure group died from cardiovascular disease.

(a) Find the proportion of men who died from cardiovascular disease in the high-blood-pressure group. Then calculate the odds.

(b) Do the same for the low-blood-pressure group.

(c) Now calculate the odds ratio with the odds for the high-blood-pressure group in the numerator. Describe the result in words.

15.2 To what extent do syntax textbooks, which analyze the structure of sentences, illustrate gender bias? A study of this question sampled sentences from ten texts. One part of the study examined the use of the words "girl," "boy," "man," and "woman." We will call the first two words juvenile and the last two adult. Here are data from one of the texts. (From Monica Macaulay and Colleen Brice, "Don't touch my projectile: gender bias and stereotyping in syntactic examples," *Language*, 73, no. 4 (1997), pp. 798–825.)

Gender	n	X (juvenile)
Female	60	48
Male	132	52

(a) Find the proportion of the female references that are juvenile. Then transform this proportion to odds.

(b) Do the same for the male references.

(c) What is the odds ratio for comparing the female references to the male references? (Put the female odds in the numerator.)

Chapter 15 Exercises

15.3 Refer to the study of cardiovascular disease and blood pressure in Exercise 15.1. Computer output for a logistic regression analysis of these data gives the estimated slope $b_1 = 0.7505$ with standard error $SE_{b_1} = 0.2578$.
  (a) Give a 95% confidence interval for the slope.
  (b) Calculate the X^2 statistic for testing the null hypothesis that the slope is zero and use Table F to find an approximate P-value.
  (c) Write a short summary of the results and conclusions.

15.4 The data from the study of gender bias in syntax textbooks given in Exercise 15.2 are analyzed using logistic regression. The estimated slope is $b_1 = 1.8171$ and its standard error is $SE_{b_1} = 0.3686$.
  (a) Give a 95% confidence interval for the slope.
  (b) Calculate the X^2 statistic for testing the null hypothesis that the slope is zero and use Table F to find an approximate P-value.
  (c) Write a short summary of the results and conclusions.

15.5 The results describing the relationship between blood pressure and cardiovascular disease are given in terms of the change in log odds in Exercise 15.3.
  (a) Transform the slope to the odds and the 95% confidence interval for the slope to a 95% confidence interval for the odds.
  (b) Write a conclusion using the odds to describe the results.

15.6 The gender bias in syntax textbooks is described in the log odds scale in Exercise 15.4.
  (a) Transform the slope to the odds and the 95% confidence interval for the slope to a 95% confidence interval for the odds.
  (b) Write a conclusion using the odds to describe the results.

15.7 To be competitive in global markets, many U.S. corporations are undertaking major reorganizations. Often these involve "downsizing" or a "reduction in force" (RIF), where substantial numbers of employees are terminated. Federal and various state laws require that employees be treated equally regardless of their age. In particular, employees over the age of 40 years are in a "protected" class, and many allegations of discrimination focus on comparing employees over 40 with their younger coworkers. Here are the data for a recent RIF:

	Over 40	
Terminated	No	Yes
Yes	7	41
No	504	765

(a) Write the logistic regression model for this problem using the log odds of a RIF as the response variable and an indicator for over and under 40 years of age as the explanatory variable.

(b) Explain the assumptions concerning binomial distributions in terms of the variables in this exercise. To what extent do you think that these assumptions are reasonable?

(c) Software gives the estimated slope $b_1 = 1.3504$ and its standard error $SE_{b_1} = 0.4130$. Transform the results to the odds scale. Summarize the results and write a short conclusion.

(d) If additional explanatory variables were available, for example, a performance evaluation, how would you use this information to study the RIF?

15.8 A study of alcohol use and deaths due to bicycle accidents collected data on a large number of fatal accidents. For each of these, the individual who died was classified according to whether or not there was a positive test for alcohol and by gender. Here are the data. (From Guohua Li and Susan P. Baker, "Alcohol in fatally injured bicyclists," *Accident Analysis and Prevention*, 26 (1994), pp. 543–548.)

Gender	n	X (tested positive)
Female	191	27
Male	1520	515

Use logistic regression to study the question of whether or not gender is related to alcohol use in people who are fatally injured in bicycle accidents.

15.9 In Examples 15.5 and 15.9, we analyzed data from the CHEESE data set described in the Data Appendix. In those examples, we used Acetic as the explanatory variable. Run the same analysis using H2S as the explanatory variable.

15.10 Refer to the previous exercise. Run the same analysis using Lactic as the explanatory variable.

15.11 For the cheese data analyzed in Examples 15.5, 15.9, and 15.10 and in the two exercises above, there are three explanatory variables. There are three different logistic regressions that include two explanatory variables. Run these. Summarize the results of these analyses, the ones using each explanatory variable alone, and all three explanatory variables together. What do you conclude?

The following four exercises use the CSDATA data set described in the Data Appendix. We examine models for relating success as measured by the GPA to several explanatory variables. In Chapter 11 we used multiple regression methods for our analysis. Here, we define an indicator variable, say HIGPA, to be 1 if the GPA is 3.0 or better and 0 otherwise.

Chapter 15 Exercises

15.12 Use a logistic regression to predict HIGPA using the three high school grade summaries as explanatory variables.

(a) Summarize the results of the hypothesis test that the coefficients for all three explanatory variables are zero.

(b) Give the coefficient for high school math grades with a 95% confidence interval. Do the same for the two other predictors in this model.

(c) Summarize your conclusions based on parts (a) and (b).

15.13 Use a logistic regression to predict HIGPA using the two SAT scores as explanatory variables.

(a) Summarize the results of the hypothesis test that the coefficients for both explanatory variables are zero.

(b) Give the coefficient for SAT math with a 95% confidence interval. Do the same for the SAT verbal score.

(c) Summarize your conclusions based on parts (a) and (b).

15.14 Run a logistic regression to predict HIGPA using the three high school grade summaries and the two SAT scores as explanatory variables. We want to produce an analysis that is similar to that done for the case study in Chapter 11.

(a) Test the null hypothesis that the coefficients of the three high school grade summaries are zero; that is, test $H_0: \beta_{HSM} = \beta_{HSS} = \beta_{HSE} = 0$.

(b) Test the null hypothesis that the coefficients of the two SAT scores are zero; that is, test $H_0: \beta_{SATM} = \beta_{SATV} = 0$.

(c) What do you conclude from the tests in (a) and (b)?

15.15 In this exercise we investigate the effect of gender on the odds of getting a high GPA.

(a) Use gender to predict HIGPA using a logistic regression. Summarize the results.

(b) Perform a logistic regression using gender and the two SAT scores to predict HIGPA. Summarize the results.

(c) Compare the results of parts (a) and (b) with respect to how gender relates to HIGPA. Summarize your conclusions.

15.16 In Example 2.32 (page 189) we studied an example of Simpson's paradox, *the reversal of the direction of a comparison or an association when data from several groups are combined to form a single group.* The data concerned two hospitals, A and B, and whether or not patients undergoing surgery died or survived. Here are the data for all patients:

	Hospital A	Hospital B
Died	63	16
Survived	2037	784
Total	2100	800

And here are the more detailed data where the patients are categorized as being in good condition or poor condition:

Good condition			Poor condition		
	Hospital A	Hospital B		Hospital A	Hospital B
Died	6	8	Died	57	8
Survived	594	592	Survived	1443	192
Total	600	600	Total	1500	200

(a) Use a logistic regression to model the odds of death with hospital as the explanatory variable. Summarize the results of your analysis and give a 95% confidence interval for the odds ratio of Hospital A relative to Hospital B.

(b) Rerun your analysis in (a) using hospital and the condition of the patient as explanatory variables. Summarize the results of your analysis and give a 95% confidence interval for the odds ratio of Hospital A relative to Hospital B.

(c) Explain Simpson's paradox in terms of your results in parts (a) and (b).

NOTES

1. Logistic regression models for the general case where there are more than two possible values for the response variable have been developed. These are considerably more complicated and are beyond the scope of our present study. For more information on logistic regression, see A. Agresti, *An Introduction to Categorical Data Analysis*, Wiley, New York, 1996; and D. W. Hosmer and S. Lemeshow, *Applied Logistic Regression*, Wiley, New York, 1989.

2. This example is taken from a classic text written by a contemporary of R. A. Fisher, the person who developed many of the fundamental ideas of statistical inference that we use today. The reference is D. J. Finney, *Probit Analysis*, Cambridge University Press, Cambridge, 1947. Although not included in our analysis, it is important to note that the experiment included a control group that received no insecticide. No aphids died in this group. We have chosen to call the response "dead." In Finney's text, the category is described as "apparently dead, moribund, or so badly affected as to be unable to walk more than a few steps." This is an early example of the need to make careful judgments when defining variables to be used in a statistical analysis. An insect that is "unable to walk more than a few steps" is unlikely to eat very much of a chrysanthemum plant!